AF730536

This book was written within the traditional homelands of the Dakota people.

We want to acknowledge the peoples on whose land we live, learn, and work.

@2023 Dan the Fish Publishing

ISBN 978-1-7352026-7-9

Contact us at:
danthefishbooks@gmail.com

Dana Digs Dinosaurs

Book Two: Meet Lynnea

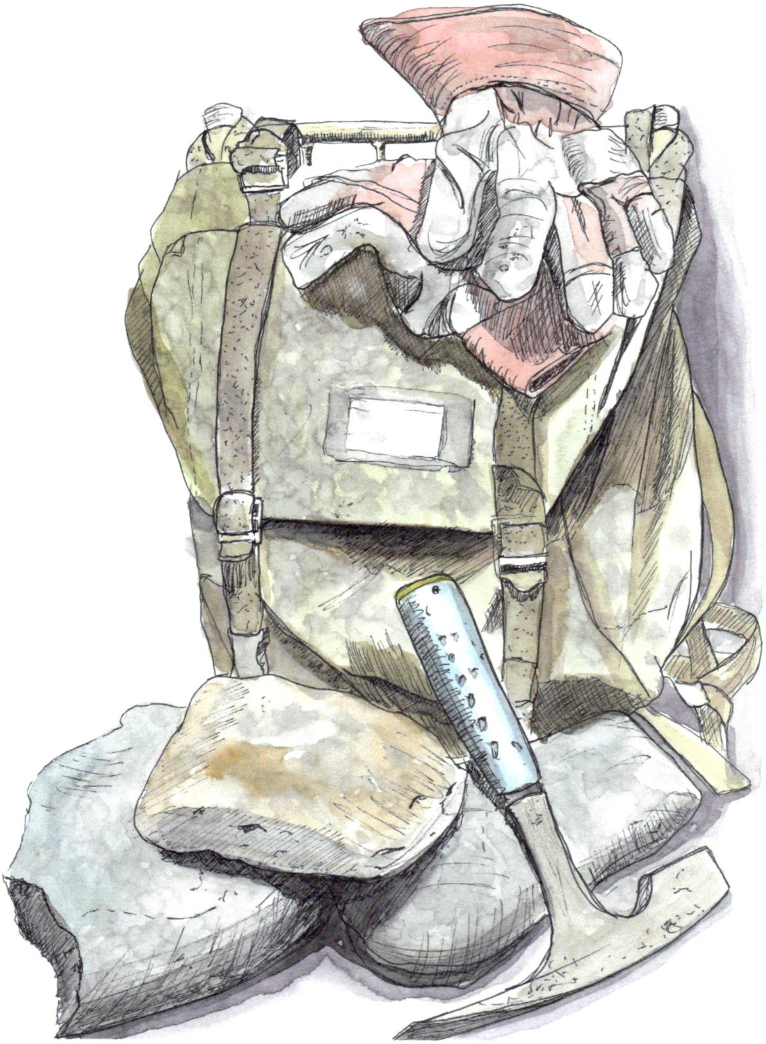

Written by Lynnea Jackson
Illustrated by Lee Brown

Hi! I’m Lynnea, the purple paleontologist.

I love everything purple.

I like to wear purple clothes, and I have purple hair.

I love purple flowers and purple cupcakes and purple rocks.

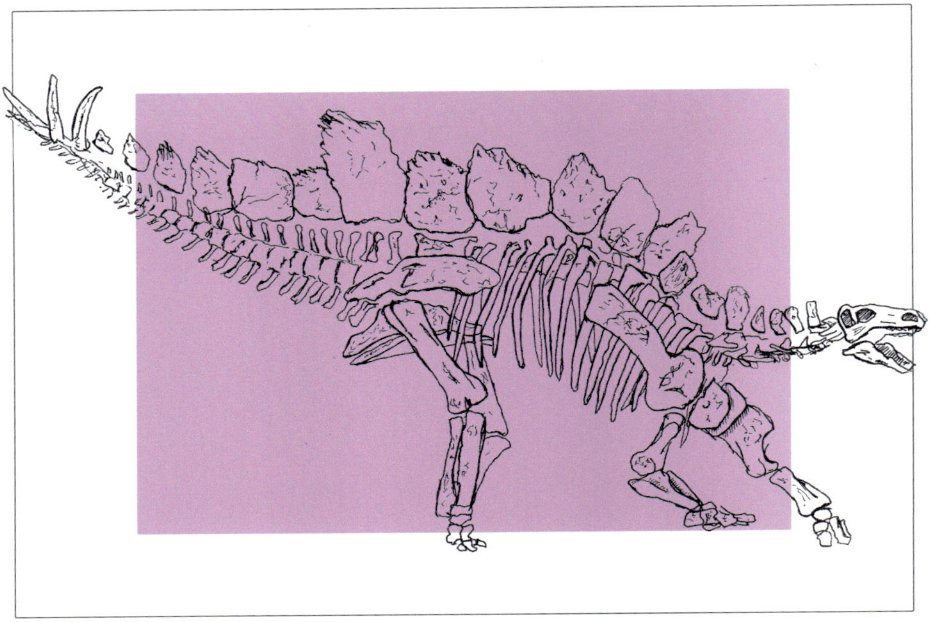

I am a paleontologist.
That means I study the history of life on Earth by looking at fossils. Fossils are the remains of animals that have been preserved as rocks.

When I was seven, I went to the Field Museum in Chicago and held a fossil in my hand.

It was then and there that I decided I wanted to be a paleontologist.

I also learned to play the cello that same year. For a while, I thought I would be a great cellist. But when I went to college, I remembered holding that fossil and chose to study paleontology.

The type of dinosaurs I am most interested in are sauropodomorphs (sor-oh-po-do-morf). Sauropods include some of the largest animals to ever walk on the Earth!

They had four legs, a long neck, and a long tail.

Even as big as they were, they were herbivores, which means they ate plants.

Paleontologists have discovered that the earliest ancestors of this type of dinosaur lived over 235 million years ago.

In my project, I'm looking at each bone of the skull of an ancient sauropodomorph from Antarctica to describe all of its features.

I make 3D models of bones using a computer-generated program. I take a CT scan with hundreds of slices of X-ray pictures of the fossil, and I study the outline of each skull bone. This gives me a way to study the bone structure and compare one fossil to another.

I travel to museums that have fossils and study the differences between species to get a better understanding of the types of animals that lived millions of years ago.

I hope my study leads to the discovery of a new species that no one has ever recorded.

One of the most interesting things about some early sauropods was that they lived in Antarctica before Antarctica was covered with ice.

During this time period, called the Early Jurassic, the world looked a lot different than what we know today.

The continents were all squished together, and the climate was different.

The remains of early sauropods have been found all over the world, in places like South Africa and South America. It shows how connected the Earth really is.

My research helps scientists understand what the world was like during the Jurassic period. The more we understand, the more we can learn about dinosaurs and their environment.

Definitions

Fossil- a trace of a once living thing from a past geologic age.

Research- investigating something to establish facts.

Environment- the natural world in which people, animals, and plants live.

Continents- a massive area of land. We currently have seven continents.

CT Scan- technology that uses a combination of X-rays to show a detailed image.

Element- tiny particles that make up matter.

Isotope- a rare version of an element.

Questions I have for you:

When you take a walk, what do you look at? Do you look up or down?

Do you like to draw dinosaurs? What medium do you use when you draw? Felicia White likes to use colored pencils when she draws dinosaurs.

drawing by Felicia White

Lynnea answers your questions

How do paleontologists know that a fossil is a million years old?

Part of being a paleontologist is understanding geology. We can learn so much about the Earth's past environments by studying the different types of rocks that have accumulated over millions of years of Earth history.

Through processes like erosion and uplift, different layers of very old rock that was once buried deep underground can be lifted to the surface where we might find fossils.

By studying the order of the layers of different types of rocks we can start to understand the relative ages that each rock layer represents.

To get an even more exact date on different layers of rock, scientists can take samples to study the isotopes in rocks. Isotopes break down at a rate that allows us to narrow down the age to a much more specific time in terms of millions of years before the present.

A fossil sauropod vertebrae being dug out of the ground.

Is the movie Jurassic Park based on scientific study, or is it just pretend?

While those movies are very entertaining, they are mostly fictional with only inspiration from real science.

The idea that DNA could be contained in mosquito blood in amber is a very exciting possibility. However, DNA deteriorates over time so it would not be usable after millions of years. So, no, we won't be cloning any T.rexes any time soon.

About the Author

Lynnea Jackson is a PhD student at the University of Michigan where she is studying vertebrate paleontology. She is originally from Okemos, Michigan. She has a Master's degree in vertebrate paleontology from the University of Minnesota and a Bachelor's degree in Geology from Michigan State University.

Lynnea loves the color purple. She loves cats. She loves painting her nails. She loves losing track of time in museums. And, she hopes that you can see yourself as a scientist, too!

About the Illustrator

My name is Lee Brown. I am an illustrator and fossil hunter living on the Jurassic Coast, South West England. In 2021, I graduated from Falmouth University with a BA in Drawing, and have since dedicated my practice to all things extinct.

I specialize in drawing dinosaurs and other prehistoric life using traditional mediums such as pen and ink, watercolor and colored pencils, among other things. I often work with museums and similar institutions to use drawing as a means to help them convey stories about the past.

Contact Lee at:

www.instagram.com/dailydinosketch

Made in United States
Troutdale, OR
11/03/2026

56844678R00019